Science

A review of inspection findings 1993/94

A report from the Office of Her Majesty's Chief Inspector of Schools

London: HMSO

Office for Standards in Education
Alexandra House
29–33 Kingsway
London WC2B 6SE

Telephone 0171-421 6800

Contents

Annex B

Introduction

This subject profile for science provides a review of findings from inspection conducted for and by OFSTED during the academic year 1993/94. It continues the publication by OFSTED of subject reports focused on the quality of provision made for and standards in science. It extends information and discussion to include aspects of inspecting science which will be of direct relevance to inspectors and may be found relevant by schools.

The evaluation of standards, quality of education and provision for science is based on evidence from the inspection of 79 primary and 735 secondary schools. The secondary schools were inspected for OFSTED by teams led by Registered Inspectors and the primary schools by teams led by Her Majesty's Inspectors of Schools (HMI) and usually containing independent inspectors in training.

In addition, evidence from the inspection of science in primary and secondary schools by science specialist HMI was used to assist in the interpretation of patterns emerging from analysis of the main body of inspection data.

Subject Report

Main Findings

- Pupils are generally gaining sound knowledge and understanding across Attainment Targets 2-4 in all Key Stages. In Key Stages 3 and 4 levels of achievement in Attainment Target 1 are lower than in the other Attainment Targets. (Paragraphs 2, 4, 6 and 11)

- Standards of achievement were satisfactory or better in around four-fifths of lessons in all Key Stages, except for Key Stage 2 where over a quarter were unsatisfactory. In the sixth form, standards were at least satisfactory in about nine out of ten lessons. (Paragraphs 1, 3, 5, 10 and 14)

- Achievement was particularly good in upper ability classes in secondary schools. (Paragraphs 5 and 10)

- Science lessons are mostly well planned, with clear objectives, but over-prescription limits achievement in a significant proportion of lessons in all Key Stages and in the sixth form. (Paragraphs 16, 19, 21, 26 and 28)

- The quality of teaching was at least satisfactory in almost four-fifths of lessons in Key Stages 3 and 4. However, teaching was unsatisfactory in over a quarter of lessons in Key Stages 1 and 2. Teaching was good or very good in almost half the sixth form lessons and unsatisfactory in only one in ten. (Paragraphs 16, 18, 20, 24 and 27)

- The majority of secondary schools have sound arrangements for National Curriculum science assessment and recording. There are unacceptably wide variations in the quality of science assessment, recording and reporting in primary schools. (Paragraphs 29 and 31)

- Procedures for the retention of evidence to support awarded levels and the standardisation of teachers' assessments are a weak feature in many primary and secondary schools. (Paragraphs 29, 31 and 33)

- There is a lack of detailed curriculum planning for science at the whole school level in primary schools which hampers the monitoring of the experience of individual pupils. (Paragraph 36)

- Departmental management of science in secondary schools is generally sound and often good. Weak management is associated with poor standards of achievement. (Paragraph 42)

Key issues for schools

Primary schools

- Steps need to be taken to enhance the science subject knowledge of teachers, especially those teaching older Key Stage 2 classes. (Paragraphs 4, 19 and 43)

- Appropriate strategies should be developed for assessing pupils' progress in scientific knowledge and skills, and for ensuring the standardisation of the judgements of different teachers. (Paragraph 29)

- Whole school curriculum planning for science needs to be more systematic to ensure that all pupils have appropriate access to the full programme of study. (Paragraph 36)

- Teaching in Key Stage 2 needs to build more effectively on pupils' experience and achievement in Key Stage 1. (Paragraph 4)

- Science co-ordinators need sufficient non-teaching time to develop their role more fully, including the monitoring of science teaching throughout the school. (Paragraph 41)

Secondary schools

- Planning and teaching in science needs to take more account of the different capabilities and achievements of pupils. (Paragraphs 23, 37 and 40)

- Teaching in Key Stage 3 needs to build more effectively on pupils' experience and achievement in Key Stage 2. (Paragraphs 23 and 37)

- The range of teaching approaches used should be reviewed to ensure that pupils are appropriately challenged to think for themselves and to take more responsibility for gathering the information they need. (Paragraphs 22 and 26)

- Appropriate strategies should be developed for ensuring the standardisation of the assessments of different teachers. (Paragraphs 31 and 33)

- Attention should be given to improving the quality of marking of pupils' written work in science. (Paragraph 34)

Standards of achievement

Key Stage 1

1 Standards of achievement in science were judged to be at least satisfactory in relation to pupils' capabilities in 79% of the Key Stage 1 lessons. Standards were good or very good in 21% of lessons.

2 At this Key Stage a good foundation of knowledge and understanding and of the basic skills of scientific enquiry is being laid. Pupils show the ability to work together on practical tasks with curiosity and interest. They are mostly well motivated and participate with enthusiasm in a range of appropriate activities, including discussion, practical work and the recording of findings. Basic scientific vocabulary is being developed and pupils are beginning to learn how to formulate questions for investigation. Many pupils are able to make comparisons and arrive at conclusions based on evidence. However, pupils with greater capabilities are not always given the opportunity to develop higher order skills such as those of prediction and hypothesising.

Key Stage 2

3 Overall standards of achievement in Key Stage 2 were slightly less satisfactory than in Key Stage 1 when judged in relation to pupils' capabilities. Standards were satisfactory or better in 73% of lessons and good or very good in 20%.

4 Many pupils build on the knowledge and skills developed in Key Stage 1 but some make only limited progress and do not achieve their full potential. In particular, some abler pupils do not make the expected progress in Years 5 and 6, for example in development of investigative skills such as prediction and evaluation. Standards are highest where full account is taken of pupils' earlier experiences, investigative work is linked to appropriate scientific knowledge, and pupils have the necessary understanding to plan their own practical work. Some teachers' understanding of particular areas of science, especially the physical

sciences, is not sufficiently well developed and this gives rise to unevenness in standards, particularly in Year 5 and Year 6.

Key Stage 3

5   Standards of achievement in science were judged to be at least satisfactory in relation to pupils' ages and abilities in 81% of lessons, and good or very good in 29%. Achievement was particularly pleasing in upper ability classes where standards were good or very good in 42% of lessons.

6   Pupils generally display at least satisfactory standards of knowledge and understanding across Attainment Targets 2–4 (Sc 2–4). Standards in Sc1 (Scientific Investigation) are mostly lower than for other Attainment Targets and show greater variation between schools. While pupils are generally able to plan simple investigations and make appropriate observations and measurements, the skills of hypothesising and of evaluating results are less well developed. Good attitudes to learning science are a positive feature in most schools. Pupils show interest in the work, co-operate well and pay due attention to safety considerations.

7   Where standards are high, pupils show secure recall of previous learning and are able to apply this to unfamiliar contexts. They are able to communicate their understanding clearly, both orally and in writing, and are encouraged to formulate ideas for themselves.

8   Most pupils are able to use simple graphical techniques, particularly bar graphs and pie charts, to represent scientific data but these are not used sufficiently to extend their understanding of scientific ideas. Skills in the use of information technology (IT) to capture, store, analyse and present information are generally underdeveloped in science.

9   For the first time in 1994 data were obtained from the statutory assessment of pupils at the end of Key Stage 3. However, only around one in five schools submitted data and the DFE considers that the results are skewed towards the higher end of the ability range. The distribution of attainment was broadly similar for both the test levels and the teacher assessed levels. Around two-thirds of pupils attained level 5 or above, just over one-third reached at least level 6 and almost one in ten reached level 7. Around one-third were at level 4 or below.

This is against the expectation that the average pupil will attain between level 5 and level 6. There were no significant differences between the performance of boys and girls.

Key Stage 4

10   Standards of achievement in science in relation to pupils' ages and abilities were judged to be good or very good in 30% of Key Stage 4 lessons and at least satisfactory in 79% – a broadly similar picture to Key Stage 3. Standards were most satisfactory in upper ability classes (43% good or very good) and least satisfactory in lower ability classes (23% good or very good).

11   Much that was said above about Key Stage 3 also applies to Key Stage 4. While pupils generally make satisfactory progress across all Attainment Targets, this is less often the case with Sc1 because of a narrow emphasis on GCSE coursework requirements rather than on systematic teaching of the essential skills and understanding required for investigative science. High standards in Sc1 are achieved where investigative activities are incorporated throughout both Key Stages and set in the context of work in one of the knowledge and under-standing Attainment Targets, and pupils are given the confidence to pursue their own ideas.

12   The achievement of higher standards in science is often restricted by over-directed teaching and by poorly targeted or insufficiently chal-lenging work. While mathematical and graphical skills are sometimes used to enhance scientific understanding, there is a tendency to give insufficient attention to this aspect of the subject.

13   This year saw a further increase in the proportion of pupils taking Double Award balanced science GCSE courses – the proportion rising from 65% to 77%. The proportion taking Single Award science fell from 17% to 12% and the proportion taking the three separate sciences fell from 5% to 3%. The proportion of the cohort entered for at least one science GCSE rose from 86% to 93% but in spite of this increased participation the average points score per science entry also rose from 3.96 to 4.37. (All data refer to maintained schools only. There is further discussion of the GCSE results in paragraphs 69–71).

Sixth Form

14　Standards of achievement in sixth form science were judged to be at least satisfactory in 91% of lessons, and good or very good in 37%. Sixth form students generally demonstrate sound subject knowledge and an ability to apply this to unfamiliar situations. They are generally able to concentrate for extended periods, ask pertinent questions and work at an appropriate pace. They often show an appreciation of how the science they are studying relates to technological applications and are able to engage in informed debate about environmental, social and ethical issues from a scientific point of view.

15　In recent years, entries for GCE A-level and AS sciences have been relatively stable compared with some other subjects. Although there has been a small reduction in the numbers taking chemistry and physics, these falls have broadly matched the demographic decline. Entries in biology have increased slowly. From 1992 to 1994, the proportion of candidates gaining grades A and B increased from 25.1% to 25.8% in biology, 30.4% to 31.5% in chemistry and 27.2% to 30.2% in physics. There have been similar increases in the proportions gaining grades A to E. (There is further discussion of the GCE results in paragraphs 73–74).

Quality of teaching

Key Stage 1

16　Teaching was judged to be satisfactory or better in 73% of the lessons and good or very good in 33%. Lessons are mostly well planned with clear learning objectives and most teachers provide a range of relevant, motivating and National Curriculum related tasks for pupils.

17　Practical work for instance often draws on pupils' own experiences, involving pupils in an active way at all stages and promoting the skills of scientific investigation (Sc1). The needs of pupils of different ability are met effectively in many cases by, for example, providing activities in which all can participate, with a variety of recording techniques available to match writing skills. Teaching is less successful when tasks and outcomes are too closely prescribed so that pupils have insufficient mental engagement with what is to be learned.

Achievement is less satisfactory when the range of teaching strategies is narrow. The over-use of worksheets or whole class instruction often leads to an overemphasis on routine recording rather than the development of scientific knowledge, understanding and skills.

Key Stage 2

18   The quality of science teaching in Key Stage 2 was slightly less satisfactory overall than in Key Stage 1. Sixty-nine per cent of the lessons were judged at least satisfactory and 27% good or very good.

19   Preparation is mostly good and teachers employ a suitable range of methods, including demonstrations and clear exposition as well as practical work of various types. Investigative activities (Sc1) are, however, often less well linked with the development of knowledge and understanding than in Key Stage 1 and opportunities for pupil investigation are fewer. Over-direction is a weakness of many of the poorer lessons. In the upper years of the Key Stage, shortcomings in teachers' understanding of science are evident in the incorrect use of scientific terminology and an overemphasis on the acquisition of knowledge at the expense of conceptual development. Pupils are at widely different stages in their scientific development by the end of the Key Stage. Planning at both class and whole school level does not take sufficient account of this.

Key Stage 3

20   The quality of teaching was judged to be at least satisfactory in 79% of lessons and good or very good in 40%. The quality of teaching to upper ability classes stands out from all other groups, with good or very good teaching in almost a half of the lessons and unsatisfactory provision in only around one in ten.

21   The teachers' command of their subject is generally very secure and lessons contain material appropriate to National Curriculum programmes of study. Most teaching is well prepared and has clear objectives, though these are not always successfully communicated to the pupils. The pace and challenge of lessons is more variable and this is often a crucial factor in determining the overall quality of the learning achieved by the pupils. Learning is most effective when pupils

understand the purpose of the work in hand, tackle challenging but accessible tasks and make effective use of prior understanding.

22   A wide and appropriate range of teaching strategies is used, particularly in the more effective science departments. These include clear exposition, class discussion, practical teacher demonstrations, class practical work, library research and the study of videotaped material. However, little use is made in the majority of schools of IT in science teaching and textbooks are frequently poorly used. Small group activities, other than for practical work, are very uncommon. In some schools, the overuse of worksheets leads to a decline in concentration and persistence, often undermining the pupils' confidence to take responsibility for their own work. Learning that is over-directed by the teacher or by worksheets often produces only superficial knowledge without real understanding.

23   Teaching seldom takes sufficient account of pupils' capabilities and previous learning. A majority of schools make use of setting arrangements, particularly in Years 8 and 9, to group together pupils of similar abilities and achievements. However, even here lesson activities and learning resources are often much the same for all classes. A minority of schools have produced clear guidance for teachers on suitable approaches for different ability groups, sometimes supported by targeted worksheets. Some of the most promising developments occur when Special Needs staff work collaboratively with science teachers in the planning and teaching of lessons.

Key Stage 4

24   The overall quality of teaching in Key Stage 4 is much the same as in Key Stage 3. Teaching is at least satisfactory in 79% of lessons and good or very good in 40%. Teaching tends to be better in upper ability classes (86% at least satisfactory, 51% good or very good) and less satisfactory in middle and lower ability groups (75% at least satisfactory and 36% good or very good).

25   All of the features of teaching in Key Stage 3 that were noted above apply also to Key Stage 4, often with even greater effect. The highest standards are achieved where: teachers have a good command of their subject, plan lessons thoroughly with clear and appropriate

objectives that are shared with the pupils, and use a range of activities in lessons that have a good pace and level of challenge that is matched to the pupils' prior knowledge and understanding; practical and investigative skills are given due attention and are systematically developed alongside the teaching of scientific facts and concepts; effective use is made of questioning to the whole class and to individuals to monitor progress and to stimulate thought. The progress of pupils is enhanced in a minority of schools by the explicit development of study skills and by involvement in well-focused discussion activities.

26   Poor standards tend to be associated with lessons that are poorly planned and ill-targeted, where the purpose is not clear to pupils, and where there is a lack of challenge and the pace is slow. These features are often linked with excessive use of copied or dictated notes, little questioning or discussion to foster interest and probe understanding, and practical work that is little more than recipe following.

Sixth Form

27   Sixth form science teaching is mostly competent. The teaching in 90% of lessons was judged at least satisfactory, and it was good or very good in 48%.

28   Particularly strong features are the quality of subject knowledge of the teachers and the good relationships apparent in the classes. The best teaching is challenging and rigorous but there is a tendency in too many classes to "spoon-feed" rather than to encourage students to think ideas through for themselves. In some cases there is over-reliance on dictated notes and duplicated handouts, and perceived examination requirements can lead to a rather narrow learning experience. As a result, students have little opportunity to check and consolidate their understanding before moving on to further ideas. However, the use in some schools of supported self-study materials encourages more independent learning and the development of valuable study skills.

Assessment, recording and reporting

Key Stages 1 and 2

29 The quality of assessment and recording procedures in primary schools is very variable, though often better at Key Stage 1 than Key Stage 2. Some schools have incorporated assessment opportunities into their curriculum planning and have appropriate systems for recording achievement, the retention of evidence and the standardisation of teachers' judgements. Many, however, still rely heavily on the unmoderated judgement of individual teachers, with recording systems indicating coverage rather than the progress of individual pupils.

30 Reporting to parents generally meets statutory requirements but is rarely well founded on good assessment and recording practice. Comments on pupil performance often lack precision and are not sufficiently diagnostic.

Key Stages 3 and 4

31 Science departments are generally meeting the statutory requirements for National Curriculum assessment, recording and reporting. However, the extent to which arrangements are effective and reliable is more variable. Weaknesses most commonly relate to incomplete records of teacher assessments or the failure to collect sufficient reliable evidence to support the levels assigned to pupils.

32 The assessment of Sc1 continues to cause difficulties for many teachers but adequate procedures are becoming established in most schools. At Key Stage 4, some schools complete only the minimum number of investigations required for GCSE moderation purposes, giving pupils insufficient opportunities to develop their skills fully.

33 The most effective record systems provide a profile of achievement and help to identify areas requiring further development. Folios of annotated work sometimes form a useful basis for discussing progress with pupils, as well as providing evidence for standardisation and moderation. Many schools have yet to put in place adequate arrangements for standardising the assessments of different teachers. Even where departments have a comprehensive system for recording

attainment, the information is rarely used to inform the planning of future teaching or to support the learning of individual pupils.

34   Weaknesses in the routine marking of pupils' work are common and include a lack of common approach, inconsistent standards, a failure to focus on the scientific content of the work, and little helpful feedback to the pupils.

Sixth Form

35   The quality of assessment post-16 is generally significantly higher than pre-16. Greater thought is given to setting a wider range of appropriate tasks and the marking is often thorough and detailed. Particularly where students are following modular courses, there is effective use of individual action plans linked to regular reviews of completed assignments.

Curriculum content

Key Stages 1 and 2

36   The majority of primary schools have science curriculum policy statements and rudimentary whole school programmes of work which take account of statutory requirements. However, documentation is often insufficiently detailed to have a major influence on teaching. Detailed planning is left to individual teachers or to year group teams who frequently present their plans in different formats, which makes whole school monitoring difficult. Science planning usually includes all attainment targets, though references to Sc1 are often too general to ensure that appropriate activities are provided for pupils.

Key Stage 3

37   Schemes of work and associated documentation mostly provide effective guidance to teachers on how to sequence the work to be covered, and on suitable teaching approaches and resources. They often helpfully indicate opportunities for assessment. They less commonly indicate appropriate ways of matching work to pupils of different levels of achievement, or seek to build effectively on pupils' learning in Key Stage 2. Schemes often pay insufficient attention to the

teaching of Sc1 and to making clear the links with the other attainment targets.

Key Stage 4

38 Curriculum planning in Key Stage 4 is mostly closely related to the requirements of the GCSE syllabus. Where the syllabus clearly indicates the different requirements for pupils aiming at different tier papers in the examination, this is sometimes reflected in better targeted work for classes of different levels of achievement.

Provision for pupils with special educational needs

Key Stages 1 and 2

39 In many primary schools good use is made of classroom ancillaries to support individual pupils and small groups, particularly the less able. This resource overcomes, to some extent, the limitations of the frequently undifferentiated planning. Some teachers provide extension tasks for the more able but there is little adaptation of worksheets or other written materials to meet the needs of the poor readers. In very few cases are those with particular scientific aptitude given specially challenging tasks.

Key Stages 3 and 4

40 The picture in ordinary schools in relation to science provision for pupils with SEN is very mixed. At best, pupils in need of support are carefully identified and targeted help provided through in-class support and modified teaching materials. Those providing in-class support, whether they be teachers, learning assistants or volunteer sixth formers, are only effective when there is a shared understanding both of the scientific objectives of the lessons and of their specific support role, and where there is continuity of contact with those receiving support. More commonly, support is erratic and often lacking focus on the specific needs of pupils in the class.

Management and administration

Primary schools

41   In a substantial minority of schools, the science co-ordinator has some non-teaching time to work alongside colleagues and to disseminate knowledge and skills gained through INSET. Where this occurs, it has a very positive effect on subject development in the school. In the majority of schools this time is not available and the influence of the co–ordinator is greatly reduced.

Secondary schools

42   Departmental management in science is generally sound and often a strength. However, there are significant weaknesses in a minority of schools and these are often those where standards of achievement are unsatisfactory.

Resources and their management

Teaching and non-teaching staff

Primary schools

43   In the great majority of schools, science is taught by the class teacher in Key Stages 1 and 2. Many primary teachers lack the depth of subject knowledge required to underpin rigorous science teaching, especially with older pupils. In a small number of schools, there is some specialist or semi-specialist teaching of science, particularly in Years 5 and 6. This can be an effective strategy for ensuring that pupils are taught by someone with a sound understanding of the subject.

Secondary schools

44   Science teachers are generally appropriately qualified and experienced, with a good match of specialisms to curriculum need. The specialist skills of staff are effectively deployed in the majority of schools. Shortages of sufficient specialist science teachers in some middle schools leads to some inaccurate and less effective teaching of science in Key Stage 3.

45   Laboratory technicians make an effective contribution to the teaching of science in the great majority of secondary schools. However, the lack of sufficient technical support in some schools sometimes leads to teachers carrying out tasks that should more efficiently be done by technicians. This weakness is particularly common in middle schools.

Resources for learning

Primary schools

46   The range of science equipment and resources is satisfactory in the majority of primary schools to meet most of the requirements of the National Curriculum. The quantity of equipment available usually dictates the mode of classroom organisation, with only sufficient available for one or two small groups to carry out the same science activity at one time.

Secondary schools

47   The effective deployment of learning resources in lessons is an important factor in the attainment of high standards of achievement in science. Strengths outweigh weaknesses in the use of learning resources in the majority of schools. There are generally sufficient equipment and materials to support the teaching of National Curriculum and sixth form courses. Shortages of some basic equipment in a minority of schools limit the range of practical work that pupils can carry out in small groups and hamper the effective development of the skills of scientific investigation. There are few schools where IT is well used to support pupils' learning in science.

48   Most pupils only have the use of a textbook in class, which restricts the range of tasks that can be set for homework. Most schools supplement texts with their own printed materials. While these often usefully support the teaching, some departments make excessive use of worksheets that provide a narrow learning diet for pupils.

Accommodation

Primary schools

49   Accommodation is satisfactory for teaching science in the majority of primary schools. However, cramped classrooms and lack of running water present difficulties for science teaching in a significant minority of schools, particularly for work in Key Stage 2.

Secondary schools

50   There are sufficient science laboratories in most secondary schools and this accommodation effectively supports the teaching. In about one-fifth of schools, however, a shortage of laboratories restricts the quality of provision in science and sometimes limits the access of pupils to the Double Science course in Key Stage 4.

Inspection issues

Inspection development

51 Inspections carried out under Section 9 of the Education (Schools) Act 1992 began in September 1993. Inspection teams have made a good start in meeting the requirements of the Framework for the Inspection of Schools; they have become more confident as the year has progressed and some early uncertainties have been resolved in many cases. This part of the report draws together some of the key issues for further improving the quality and standard of inspection. Many issues are similar from one subject to another; where there are subject-specific matters these are indicated.

52 Some examples of inspection writing are included. They are not intended to be viewed as models or templates but illustrate how some inspectors have more effectively completed forms and met Framework requirements.

Evidence gathering

53 Inspectors generally sample a good range of science work of different year groups, abilities and key stages across the compulsory years of education. They usually achieve a good balance although the time allocated to inspecting science varies considerably. It is important that where a school has a sixth form, post-16 work is fairly represented in the sampling.

54 In reaching their judgements, inspectors use evidence from a good range of sources. It is important that clear reference is made to them in writing to support judgements. The Supplementary Evidence Form provides a means of documenting evidence and judgements from sources other than lessons and could be more widely used.

Lesson Observation Forms

55 Overall, Lesson Observation Forms are completed conscientiously, with attention to the relevant evaluation criteria. Inspectors could usefully check that on these forms and in other writing subject- specific

character and detail is included wherever possible and that the match is good between grades and text.

56   In relation to the content of lessons, the majority of inspectors adequately indicate the topic of lessons usually with reference to the National Curriculum Order. Further details of the lesson activities would be helpful in setting the context. For example:

Year 6, mixed ability

Respiratory system. Recap on previous work on breathing – whole class Q and A. Demonstration of dissection of a pig's lungs – glass tube used to inflate lungs. Explanation of structure and function of lungs. Work aimed at Sc2/5a.

57   Inspectors draw on their professional knowledge and experience to make overall judgements about the **achievements** of pupils. Responding to the Framework requirements to assess pupils' achievements in relation to what are now termed national expectations and taking account of pupils' abilities has not proved easy. Revised requirements and guidance published in June 1994 were to help inspectors in making these distinct judgements. To support judgements it is important that inspectors clearly identify and record what pupils know, understand and can do and set achievements in the context of National Curriculum expectations. For example:

Year 10, upper ability (class experiment on Ohm's Law and measuring resistance)

Achievement (age referenced): work up to Level 9 in Sc4. Pupils all able to plot V/I graph for resistance wire and use these skills in tackling the relationship for a bulb. Good level of practical skills and graphicacy skills. Pupils able to use ideas of ions and electrons in their explanations. Grade: 2

Achievement (taking account of pupils' abilities): Scientific understanding at least sound for capabilities, some good. Pupils being challenged appropriately. Grade: 2

58   Clear evidence of pupils' attitudes to learning and their behaviour in lessons is usually given, and this is reflected in the grade given for **quality of learning**. Greater prominence should be given to other

attributes of learning, particularly pupils' progress in lessons and the development of the science skills included in section 6.1 of Part 4 of the *Handbook for the Inspection of schools*. Care is needed to ensure that this part of the Lesson Observation Form does not include evidence and judgements more pertinent to the quality of teaching.

Year 10 (working on the classification of living things)

Pupils learned to use a key but learned nothing about the organisms involved or about the characteristics of the groups to which they belong. The pupils did not work consistently and were not aware of the purpose of the lesson. Little enthusiasm shown. Reasonable observational skills. Able to follow instructions. Grade: 4

59   Inspectors usually cite relevant evidence when judging the **quality of teaching**, and evaluation is based on the criteria in the Framework. They need to check that the full range of criteria is used, including teachers' command of the subject. In the following examples, a number of attributes of teaching are referred to.

R/Year 1, mixed ability

Clear objectives – well planned/well organised lesson. Very good range of well thought-out activities to develop pupils' ideas about their senses. Two classroom assistants well briefed and work effectively with pupils – also set out the resources while class teacher introduces the lesson. Effective interventions from teacher to probe understanding and give support to pupils' learning. Grade: 1

Year 6, mixed ability (lesson on water and changes of state)

Planning not sufficiently clear. Input by teacher unclear and insufficient sense of purpose – exposition lacks clarity. Teacher shows lack of subject knowledge which constrains the teaching. Investigation planning sheets used but not followed up – no consolidation. Lack of differentiation affects standards of least able. Marking of books unhelpful. Sound relationships ensure satisfactory class control. Grade: 4

60   The Lesson Observation Form could be more widely used to indicate contributions made by the subject lesson to key skills and to learning in other areas of the curriculum. It also provides opportunity to

signal the impact of contributory factors on achievements and the quality of learning which can be drawn on when compiling the Subject Evidence Form.

Subject Evidence Forms

61 Subject evidence forms are usually fully completed, very often thoroughly and thoughtfully. In most cases, a wide range of evidence appears to have been used. Inspectors need to check that this is sufficiently explicit in the relevant sections of the Form and to ensure that the emphasis is towards evaluation rather than description.

62 Particular attention is given to aspects of standards of achievement and the quality of learning and teaching although, as in lesson observation forms, in considering the quality of learning more emphasis is placed on pupils' attitudes and behaviour than on their skills as learners. When commenting on examination results as part of their evaluation of standards of achievement, inspectors should ensure that the evidence includes the basis for any comparisons with national data, such as the grade ranges being compared, types of school, nature of courses followed.

63 Completed sections on standards of achievement and the quality of teaching and learning from Subject Evidence Forms follow. While brief, they provide adequate evidence on which to base the subject paragraph for the report.

Standards of Achievement

Standards in relation to national expectations across the three knowledge and understanding attainment targets (Sc2,3,4) were good or better in over 60% of lessons. Evidence from lessons and scrutiny of work indicates standards in these ATs are high in the top sets in both Key Stage 3 and Key Stage 4. Pupils are generally able to talk confidently and use scientific ideas and terminology accurately. Written work is good or better across all ages and abilities: it is well presented and is accurate in content. Standards overall are consistent with pupils' capabilities. Only one lesson was seen where pupils were clearly underachieving due to a poor match between demands of the task and the pupils' capabilities.

Standards in basic practical skills of observing, measuring and recording are generally good. Investigative skills (Sc1) are generally satisfactory in both Key Stages 3 and 4, with the best work seen in Y8.

Examination results at GCSE are good. For GCSE biology and physics the percentages of grades A and B are well above the national norms. GCSE results for chemistry are satisfactory. Results for pupils taking the double award science course are good when allowance is made for the more able pupils who take the three separate sciences (approximately the top 15% of the year group).

Quality of Learning

Pupils were willing, almost always co-operative and well behaved, and in general spent the majority of time on task. Small group sizes for lower ability groups helped by providing the teacher with more opportunities for individual attention.

Pupils in both key stages generally made satisfactory progress in their learning of scientific concepts across ATs 2–4. Progress was more variable in KS3 than KS4.

Pupils' practical skills in both key stages were well developed with pupils invariably working safely. Pupils' ability to identify patterns, interpret and plot graphs was variable throughout both key stages – for a number of pupils poor graphical skills and in some cases poor numerical skills prevented effective learning. Poorly developed investigative skills also limited attainment in a number of cases – in particular, pupils' ability to hypothesise effectively making use of scientific knowledge was limited, thus denying access to higher levels in Sc1.

Quality of Teaching

The quality of teaching was satisfactory or better in 83% of lessons seen in KS3. Relationships with most pupils were very good. Lessons were well planned and resourced, pupils knew what was expected of them and were encouraged to work independently. In a small minority of lessons, the teacher found it difficult to manage the class as a whole because isolated individuals needed a disproportionate share of the teacher's attention.

The quality of teaching in all lessons at KS4 was satisfactory or better. Moreover, in 66% of these the teaching was good or very good. Relationships with most pupils were good, teachers had high expectations and gave appropriate support and encouragement. Attention was drawn to safety matters before practical work was undertaken. Some lessons were particularly fast moving. Where classes contained small numbers of less motivated pupils, teachers worked hard at establishing relationships and maintained high expectations of pupil behaviour. Teachers showed a clear understanding of the needs of different groups and managed the learning accordingly.

64   When considering features such as the resources for learning, organisation, management and procedures and, indeed, the quality of teaching, the shift of evaluation should be towards the effects on the standards achieved and the quality of learning. The following extracts from the 'Contributory Factors' sections of the Subject Evidence Form include clear judgements about features of provision and some indications of their effects.

Accommodation

Most laboratories are light, in a reasonable state of repair and decoration and able to cope with the size of classes. There is, however, a shortage of specialist rooms properly resourced to cater for the demands across the National Curriculum. This entails classes taking place in areas outside the science block with staff having to transport equipment etc. These rooms are not suitable for good science work to take place and, inevitably, cause loss of important teaching time and result in some less effective lessons. Good use is made of the specialist accommodation available with rooms being enhanced by displays of pupils' work.

Non-teaching staff

The three technicians provide outstanding support to the teachers despite the difficulty of access to over-used laboratories. Their high level of efficiency and dedication ensures that a high level of practical work can take place with a minimum time taken from the teachers' contact with their classes.

Management and administration of the subject

The department has a clear development plan and the allocation of capitation is related to the priorities in it. All members of the department are involved in the on-going review and revision of curricular provision. The head of department and KS3 co-ordinator have been instrumental in the introduction of a range of new initiatives during the past 18 months. They have succeeded in developing an excellent team spirit within the department. All members seem to share a common vision and enthusiasm, and support within the department is commendable. This is reflected in the high quality of teaching observed in most lessons and in the evidence of improving standards of achievement.

Judgement Recording Statements

65   The proformas of Judgement Recording Statements are usually fully completed. Inspectors need to ensure that all available evidence is considered in arriving at judgements for inclusion in the proforma. The purpose and use of Judgement Recording Statements are outlined in Appendix C of Part 3 of the *Handbook for the Inspection of Schools*.

Subject sections in inspection reports

66   Most science subject sections in the inspection reports scrutinised meet the Framework requirements and match the evidence in the Subject Evidence Forms. They give appropriate emphasis to standards of achievement and the quality of learning and teaching. Inspectors need to ensure that overall judgements are clear and succinct and draw on all the evidence available, and that factors which impact on standards of achievement and quality of learning are clearly identified.

67   The following extracts from four reports illustrate writing about standards of achievement, quality of teaching and learning and a contributory factor. The characteristics of science are evident.

School A

In Key Stages 3 and 4 pupils' standards of achievement in science are below average or low in relation to national expectations and are unsatisfactory in relation to pupils' capabilities. GCSE results are significantly below local and national norms. Pupils generally show very limited knowledge and understanding of scientific concepts. They are given little opportunity to use or develop their scientific skills, which are poorly developed. Pupils are not used to planning, carrying out or evaluating scientific investigations. Standards of literacy and numeracy are generally poor and significantly affect the standards of achievement in science......

School B

The progress of pupils' learning in science is generally satisfactory. Most pupils are gaining an adequate understanding of key scientific ideas. Pupils' attitudes to learning are generally good. They are willing to speculate and to look for patterns; they are inquisitive and able to co-operate in group work. They show care and respect for each other and for living things in general. They also tend to work safely and sensibly when handling scientific materials and equipment. Their skills in communicating information and ideas are well developed, particularly with numerical data, as are their problem solving skills.

School C

In the majority of lessons teachers have clear objectives, and appropriate activities enable pupils to progress at a suitable pace. Teacher expositions to the whole class engage the pupils and encourage the development of a technical vocabulary. There are, however, few opportunities for pupils to consider the practical applications and social implications of science. Relationships in the classroom are very good though there is sometimes insufficient interaction between the teacher and individuals or groups. Teachers have appropriate subject knowledge and have high expectations of pupils.

68   In writing to the Framework requirements, inspectors need to check that a comment is included on compliance with statutory requirements and that key issues for action in the subject are clearly given. These are helpful to schools in their action planning.

Interpretation of subject performance data

(Note: All data in this section relate to candidates from maintained schools only.)

GCSE

69   Care is needed in interpreting GCSE Science examination data because of the variety of course options: Single Award, Double Award and the three separate sciences. In 1994, the percentage of the Y11 cohort (477,460) entered for each of these options was:

Single Award Science   12.0%

Double Award Science   77.3%

Biology   3.1%

Chemistry   3.0%

Physics   3.0%

70   Around 93% of the cohort were entered for at least one science GCSE examination; this is broadly similar to English and Mathematics. The average points score per entry was 4.37, compared to 4.59 for English and 4.09 for Mathematics.

71   There are important differences in the nature of candidates taking each of the science options and in the gender ratios:

- **Single Award courses** tend to be taken by less able pupils (only 14.2% of candidates gained A–C grades in 1994). Slightly more girls than boys take these courses and the girls tend to achieve higher grades (16.4% A–C against 11.7% for boys).

- **Double Award courses** are taken by a wide cross-section of pupils, and about equally by boys and girls. Results from boys and girls are similar; around 45% gained A–C grades in 1994.

- **Separate Sciences** are usually taken by the more able pupils and about 33% of entries are from selective schools. Well over 70% of candidates gained grades A–C in 1994. Almost twice as many boys as girls took the separate sciences in 1994. While there was little difference in the performance of girls and boys in biology and chemistry, boys continue to out-perform girls in physics.

72 Inspectors should note the proportion of the cohort entered for GCSE science and the distribution across different types of courses. When making judgements about examination results, inspectors should bear in mind the following points:

- Comparisons with other core subjects are complicated where candidates take a variety of science courses. The mean points score per science subject entry is probably the most useful comparator.

- Care is needed in making year on year comparisons because of curriculum changes and changes in the proportions of candidates taking different routes. In the past two years there has been a significant increase in the numbers taking Double Award courses and a parallel fall in the numbers taking separate sciences. Numbers taking Single Award courses fell in 1994. Apart from Single Award courses, there have been increases in the proportions of candidates gaining grades A–C in science GCSE subjects between 1992 and 1994; this has been most marked in the case of the separate science examinations.

GCE A/AS

73 A feature of GCE A-level entries is the gender ratio for different science subjects. The ratio of boys to girls is 0.6:1 in biology, 1.4:1 in chemistry and 3.9:1 in physics. Boys tend to achieve slightly better than girls in chemistry and physics, while this is reversed in biology. Overall, the results for chemistry and physics are close to the 'all subject' averages, though somewhat below for biology.

74 GCE AS examination entries in science remain very low and any comparisons with national statistics should be made with great caution. Only around a half of candidates in biology and chemistry obtained a graded result.

75 The increasing popularity of modular courses in the sciences at A/AS level adds a new complication to the interpretation of examination statistics as candidates may defer submitting for certification for up to four years. This is likely to reduce the proportion of ungraded candidates in these examinations

Annex A

GCSE results for 15 year olds[1] for Biology 1994

Type of School		Number of 15 year old pupils entered	1994 Percentages achieving grades									1994 Average points score[3]	1994 % A*-C grades	1994 % A*-G grades	1993 Average points score[3]	1993 % A-C grades	1992[2] Average points score[3]	1992[2] % A-C grades
			A*	A	B	C	D	E	F	G	U							
Comprehensive		9950	2.9	16.8	26.5	27.1	14.4	6.8	3.2	1.3	0.4	5.26	73.3	99.0	4.37	53.7	4.18	47.4
Selective		4737	7.2	27.4	31.8	24.0	7.0	2.0	0.3	0.1	0.0	5.96	90.4	99.9	5.71	88.0	5.61	85.2
Modern		19	–	–	–	–	–	–	–	–	–	–	–	–	3.65	35.5	3.19	22.4
Maintained	All pupils	14706	4.3	20.2	28.2	26.1	12.0	5.3	2.3	0.9	0.3	5.49	78.7	99.2	4.64	60.7	4.35	52.0
	Boys	9265	4.0	19.0	28.2	27.0	12.7	5.3	2.2	0.9	0.3	5.45	78.2	99.3	4.79	64.7	4.51	56.5
	Girls	5441	4.9	22.1	28.2	24.4	11.0	5.2	2.4	1.0	0.3	5.55	79.6	99.2	4.50	57.1	4.24	48.9
All Subjects Maintained	All pupils		2.1	8.4	16.4	20.5	18.9	14.5	10.2	4.5	1.5	4.40	47.4	95.5	4.12	46.3	4.14	45.0

1 Aged 15 on 31/8/93
2 1992 results include a small amount of data from special schools
3 Calculated on basis A*=8, A=7, B=6, C=5, D=4, E=3, F=2, G=1

– less than 100 candidates
* more than 100 and less than 500 candidates
x information not available

29

GCSE results for 15 year olds¹ for Chemistry 1994

Type of School		Number of 15 year old pupils entered	1994									Average points score³	% A*-C grades	% A*-G grades	1993		1992²	
			Percentages achieving grades												Average points score³	% A-C grades	Average points score³	% A-C grades
			A*	A	B	C	D	E	F	G	U							
Comprehensive		9643	4.0	15.6	24.2	24.9	14.9	9.4	4.6	1.4	0.4	5.14	68.6	99.0	4.49	56.9	4.34	53.5
Selective		4705	8.7	25.2	29.1	22.8	9.5	3.7	0.7	0.1	0.1	5.86	85.8	99.8	5.72	84.7	5.66	84.6
Modern		22	–	–	–	–	–	–	–	–	–	–	–	–	3.09	29.5	3.26	23.4
Maintained	All pupils	14370	5.5	18.7	25.7	24.2	13.1	7.6	3.4	1.0	0.3	5.38	74.2	99.3	4.75	62.8	4.51	57.4
	Boys	9193	5.2	18.2	26.0	25.1	13.3	7.2	3.2	1.0	0.4	5.37	74.5	99.2	4.82	64.4	4.56	58.9
	Girls	5177	6.1	19.6	25.3	22.7	12.7	8.3	3.6	1.0	0.1	5.39	73.7	99.3	4.65	60.6	4.44	55.5
All Subjects Maintained	All pupils		2.1	8.4	16.4	20.5	18.9	14.5	10.2	4.5	1.5	4.40	47.4	95.5	4.12	46.3	4.14	45.0

1 Aged 15 on 31/8/93

2 1992 results include a small amount of data from special schools

3 Calculated on basis A*=8, A=7, B=6, C=5, D=4, E=3, F=2, G=1

– less than 100 candidates

* more than 100 and less than 500 candidates

x information not available

GCSE results for 15 year olds[1] for Physics 1994

Type of School		Number of 15 year old pupils entered	Percentages achieving grades									Average points score[3]	% A*-C grades	% A*-G grades	Average points score[3]	% A-C grades	Average points score[3]	% A-C grades
												1994			**1993**		**1992[2]**	
			A*	A	B	C	D	E	F	G	U							
Comprehensive		9695	5.4	16.6	21.1	25.4	15.2	9.6	4.1	1.5	0.6	5.18	68.4	98.8	4.60	57.9	4.45	54.1
Selective		4734	12.8	25.4	26.8	22.6	9.1	2.4	0.7	0.0	0.0	6.00	87.6	99.8	5.84	87.0	5.76	86.7
Modern		38	–	–	–	–	–	–	–	–	–	–	–	–	3.49	31.3	3.22	25.1
Maintained	All pupils	14467	7.8	19.4	22.9	24.5	13.2	7.3	3.0	1.1	0.4	5.44	74.6	99.1	4.88	64.2	4.61	58.1
	Boys	9328	9.2	20.4	23.4	23.9	12.4	6.4	2.6	0.9	0.4	5.56	76.9	99.2	4.85	63.5	4.54	56.5
	Girls	5139	5.3	17.6	22.1	25.5	14.5	9.0	3.8	1.4	0.5	5.24	70.4	99.0	4.94	65.9	4.77	61.9
All Subjects Maintained	All pupils		2.1	8.4	16.4	20.5	18.9	14.5	10.2	4.5	1.5	4.40	47.4	95.5	4.12	46.3	4.14	45.0

1 Aged 15 on 31/8/93
2 1992 results include a small amount of data from special schools
3 Calculated on basis A*=8, A=7, B=6, C=5, D=4, E=3, F=2, G=1

– less than 100 candidates
* more than 100 and less than 500 candidates
x information not available

GCSE results for 15 year olds¹ for Combined Science – Single Award 1994

Type of School		Number of 15¹ year old pupils entered	1994									Average points score³	% A*-C grades	% A*-G grades	1993		1992²	
			Percentages achieving grades												Average points score³	% A-C grades	Average points score³	% A-C grades
			A*	A	B	C	D	E	F	G	U							
Comprehensive		51656	0.1	0.6	4.7	8.1	24.2	23.7	17.2	8.1	5.3	3.28	13.5	86.6	2.73	16.1	2.96	18.7
Selective		780	3.5	10.1	30.1	20.8	15.3	8.6	4.0	2.1	2.9	5.09	64.5	94.4	5.20	75.0	4.65	63.9
Modern		4894	0.0	0.2	5.0	8.1	31.4	25.0	15.1	5.6	3.6	3.41	13.2	90.3	2.93	18.2	3.12	20.0
Maintained	All pupils	57330	0.1	0.7	5.1	8.3	24.7	23.6	16.9	7.8	5.1	3.31	14.2	87.0	2.76	16.6	2.99	19.1
	Boys	27025	0.1	0.5	4.0	7.0	23.9	23.7	17.6	8.4	5.8	3.21	11.7	85.2	2.61	13.8	2.84	16.4
	Girls	30305	0.1	0.9	6.0	9.4	25.4	23.4	16.2	7.3	4.6	3.40	16.4	88.6	2.90	19.1	3.12	21.5
All Subjects Maintained	All pupils		2.1	8.4	16.4	20.5	18.9	14.5	10.2	4.5	1.5	4.40	47.4	95.5	4.12	46.3	4.14	45.0

1 Aged 15 on 31/8/93

2 1992 results include a small amount of data from special schools

3 Calculated on basis A*=8, A=7, B=6, C=5, D=4, E=3, F=2, G=1

– less than 100 candidates

* more than 100 and less than 500 candidates

x information not available

GCSE results for 15 year olds[1] for Combined Science – Double Award 1994

Type of School		Number of 15[1] year old pupils entered	Percentages achieving grades									Average points score[3]	% A*–C grades	% A*–G grades	Average points score[3]	% A–C grades	Average points score[3]	% A–C grades
												1994			1993		1992[2]	
			A*	A	B	C	D	E	F	G	U							
Comprehensive		347906	2.3	6.3	16.8	18.4	22.9	16.3	9.3	3.2	1.3	4.37	43.9	95.6	4.10	44.6	×	×
Selective		10923	11.7	22.2	37.6	17.9	7.3	2.0	0.7	0.2	0.1	6.03	89.3	99.5	5.68	87.2	×	×
Modern		10086	0.3	2.1	12.8	16.2	27.0	19.8	12.7	4.0	1.8	3.91	31.3	94.9	3.71	32.6	×	×
Maintained	All pupils	368915	2.6	6.6	17.3	18.4	22.5	16.0	9.2	3.2	1.3	4.41	44.9	95.7	4.14	45.5	×	×
	Boys	185806	2.7	6.7	17.3	18.4	22.6	15.7	9.0	3.0	1.4	4.43	45.0	95.4	4.13	45.6	×	×
	Girls	183109	2.5	6.6	17.3	18.4	22.4	16.2	9.4	3.3	1.2	4.39	44.8	96.0	4.14	45.5	×	×
All Subjects Maintained	All pupils		2.1	8.4	16.4	20.5	18.9	14.5	10.2	4.5	1.5	4.40	47.4	95.5	4.12	46.3	4.14	45.0

1 Aged 15 on 31/8/93
2 1992 results include a small amount of data from special schools
3 Calculated on basis A*=8, A=7, B=6, C=5, D=4, E=3, F=2, G=

– less than 100 candidates
* more than 100 and less than 500 candidates
× information not available

Annex B

GCE AS results for Biology 1994

| | | Number of candidates | Percentages achieving grades | | | | | | | % A–B grades | % A–E grades | Average points score[p] | 1993 % A–B grades | 1993 % A–E grades | 1992 % A–B grades | 1992 % A–E grades |
			A	B	C	D	E	N	U							
Type of School																
Maintained	All pupils	1113	6.5	6.8	8.4	12.1	20.6	16.9	26.1	13.3	54.4	1.3	15.8	53.8	13.1	57.7
	Boys	404	5.4	6.9	7.2	10.1	23.5	17.8	26.2	12.4	53.2	1.2	14.1	50.3	14.2	56.8
	Girls	709	7.1	6.8	9.0	13.3	18.9	16.4	26.0	13.8	55.0	1.3	16.7	55.4	12.6	58.1
All subjects																
Maintained	All pupils		7.1	10.2	14.8	17.9	18.2	12.9	15.1	17.3	68.2	1.8	17.0	65.5	16.6	65.4

– less than 100 candidates

* more than 100 and less than 500 candidates

p Calculated on basis A=5, B=4, C=3, D=2, E=1

The number of pupils taking AS levels is insufficient to yield a meaningful analysis by type of maintained school

GCE AS results for Chemistry* 1994

					1994								1993		1992	
	Number of candidates		Percentages achieving grades						% A–B grades	% A–E grades	Average points score[p]		% A–B grades	% A–E grades	% A–B grades	% A–E grades
Type of School		A	B	C	D	E	N	U								
Maintained All pupils	486	5.1	9.3	9.9	16.7	16.9	17.1	21.6	14.4	57.8	1.4		12.3	49.6	15.6	63.4
Boys	293	5.1	8.9	10.2	14.0	20.1	18.1	20.5	14.0	58.4	1.4		13.4	52.3	14.4	63.4
Girls	193	5.2	9.8	9.3	20.7	11.9	15.5	23.3	15.0	57.0	1.5		11.1	46.3	17.3	63.5
All subjects Maintained All pupils		7.1	10.2	14.8	17.9	18.2	12.9	15.1	17.3	68.2	1.8		17.0	65.5	16.6	65.4

– less than 100 candidates

* more than 100 and less than 500 candidates

p Calculated on basis A=5, B=4, C=3, D=2, E=1

The number of pupils taking AS levels is insufficient to yield a meaningful analysis by type of maintained school

35

GCE AS results for Physics 1994

Type of School		Number of candidates	Percentages achieving grades							% A–B grades	% A–E grades	Average points score[p]	1993 % A–B grades	1993 % A–E grades	1992 % A–B grades	1992 % A–E grades
			A	B	C	D	E	N	U							
Maintained	All pupils	815	7.4	8.0	12.6	13.5	23.8	16.9	15.1	15.3	65.3	1.6	16.9	63.5	18.0	67.7
	Boys	656	8.2	7.5	11.4	12.8	24.7	17.5	15.2	15.7	64.6	1.6	16.1	62.4	17.8	68.2
	Girls	159	3.8	10.1	17.6	16.4	20.1	14.5	14.5	13.8	67.9	1.6	19.9	68.0	18.9	66.0
All subjects																
Maintained	All pupils		7.1	10.2	14.8	17.9	18.2	12.9	15.1	17.3	68.2	1.8	17.0	65.5	16.6	65.4

– less than 100 candidates

* more than 100 and less than 500 candidates

p Calculated on basis A=5, B=4, C=3, D=2, E=1

The number of pupils taking AS levels is insufficient to yield a meaningful analysis by type of maintained school

GCE A-Level results for Biology 1994

Type of School		Number of candidates	Percentages achieving grades							% A–B grades	% A–E grades	1993 % A–B grades	1993 % A–E grades	1992 % A–B grades	1992 % A–E grades
			A	B	C	D	E	N	U						
Comprehensive		16597	9.8	13.2	16.5	17.9	18.2	13.2	10.6	22.9	75.5	22.5	73.9	23.1	75.1
Selective		3489	20.3	19.5	19.9	16.1	13.2	6.7	4.0	39.8	88.9	41.9	88.8	38.4	87.5
Modern		149	3.4	6.0	10.7	16.8	20.1	16.8	20.8	9.4	57.0	8.3	50.9	18.9	68.0
Maintained	All pupils	20235	11.5	14.2	17.0	17.6	17.3	12.1	9.5	25.8	77.7	25.9	76.4	25.1	76.6
	Boys	7836	10.9	13.3	16.8	17.8	18.2	12.1	10.1	24.2	77.1	25.6	75.9	25.6	76.9
	Girls	12399	12.0	14.8	17.2	17.4	16.7	12.0	9.2	26.7	78.0	26.0	76.6	24.8	76.5
All subjects Maintained	All pupils		13.1	16.2	18.5	18.9	15.2	9.4	7.5	29.3	81.9	28.0	79.7	26.4	78.6

– less than 100 candidates

* more than 100 and less than 500 candidates

GCE A-Level results for Chemistry 1994

Type of School		Number of candidates	A	B	C	D	E	N	U	1994 % A–B grades	1994 % A–E grades	1993 % A–B grades	1993 % A–E grades	1992 % A–B grades	1992 % A–E grades
			Percentages achieving grades												
Comprehensive		13615	12.4	16.0	15.9	16.7	15.9	11.4	11.2	28.4	76.9	27.9	75.8	28.1	76.4
Selective		3097	23.9	22.0	17.1	14.9	10.9	7.2	3.6	45.9	88.7	45.2	88.4	44.3	89.5
Modern		75	–	–	–	–	–	–	–	–	–	–	–	21.5	70.1
Maintained	All pupils	16787	14.4	17.1	16.1	16.3	15.0	10.7	9.8	31.5	79.0	31.1	78.0	30.4	78.2
	Boys	9704	15.2	17.0	15.9	16.0	14.5	10.5	10.3	32.2	78.6	31.8	77.8	31.2	78.1
	Girls	7083	13.3	17.3	16.2	16.8	15.7	10.9	9.2	30.6	79.4	30.0	78.4	29.3	78.3
All subjects Maintained	All pupils		13.1	16.2	18.5	18.9	15.2	9.4	7.5	29.3	81.9	28.0	79.7	26.4	78.6

– less than 100 candidates

* more than 100 and less than 500 candidates

GCE A-Level results for Physics 1994

Type of School		Number of candidates	Percentages achieving grades 1994							% A–B grades	% A–E grades	1993 % A–B grades	% A–E grades	1992 % A–B grades	% A–E grades
			A	B	C	D	E	N	U						
Comprehensive		12385	13.5	13.8	16.0	18.1	17.2	12.1	8.8	27.3	78.6	26.9	76.3	25.1	74.9
Selective		2926	23.9	19.4	18.4	15.6	12.7	6.9	2.8	43.3	90.0	40.0	88.9	39.0	88.4
Modern		106	2.8	9.4	14.2	15.1	22.6	13.2	21.7	12.3	64.2	15.0	52.3	21.1	65.1
Maintained	All pupils	15417	15.4	14.9	16.5	17.6	16.4	11.1	7.7	30.2	80.6	29.2	78.5	27.2	76.8
	Boys	12301	16.0	14.7	16.3	17.4	16.3	11.1	7.8	30.8	80.7	29.7	78.5	27.5	76.5
	Girls	3116	12.8	15.3	17.3	18.4	16.8	11.2	7.4	28.1	80.6	27.4	78.7	25.8	77.8
All subjects Maintained	All pupils		13.1	16.2	18.5	18.9	15.2	9.4	7.5	29.3	81.9	28.0	79.7	26.4	78.6

– less than 100 candidates

* more than 100 and less than 500 candidates

Printed in the United Kingdom for HMSO
Dd300292 4/95 C130 G3397 10170